AF882727

This book belongs to:

Made in the USA
Monee, IL
08 November 2026

67972912R00083